Developing Numeracy

MENTAL MATHS

ACTIVITIES FOR THE DAILY MATHS LESSON

year R

Hilary Koll and Steve Mills

A & C BLACK

Contents

Resources

Published 2004 by A & C Black Publishers Limited
37 Soho Square, London W1D 3QZ
www.acblack.com

ISBN 0-7136-6909-8

Copyright text © Hilary Koll and Steve Mills, 2004
Copyright illustrations © Michael Evans, 2004
Copyright cover illustration © Charlotte Hard, 2004
Editors: Lynne Williamson and Marie Lister

The authors and publishers would like to thank Jane McNeill and Catherine Yemm for their advice in producing this series of books.

A CIP catalogue record for this book is available from the British Library.

Printed and bound in Great Britain by Cromwell Press Ltd, Trowbridge.

A & C Black uses paper produced with elemental chlorine-free pulp, harvested from managed sustainable forests.

Introduction

Developing Numeracy: Mental Maths is a series of seven photocopiable activity books designed to be used during the daily maths lesson. This book focuses on the skills and concepts for mental maths outlined in the National Numeracy Strategy *Framework for teaching mathematics* for Reception (also now known as Foundation 1). The activities are intended to be used in the time allocated to pupil activities; they aim to reinforce the knowledge and develop the facts, skills and understanding explored during the main part of the lesson. They provide practice and consolidation of the objectives contained in the framework document.

Mental Maths Year R

To calculate mentally with confidence, it is necessary to understand the three main aspects of numeracy shown in the diagram below. These underpin the teaching of specific mental calculation strategies.

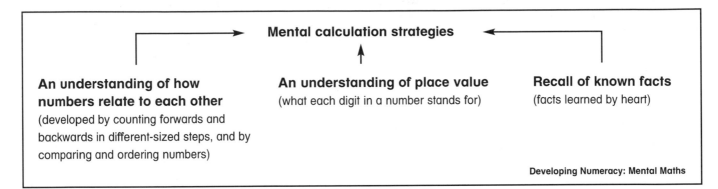

Mental calculation strategies

An understanding of how numbers relate to each other
(developed by counting forwards and backwards in different-sized steps, and by comparing and ordering numbers)

An understanding of place value
(what each digit in a number stands for)

Recall of known facts
(facts learned by heart)

Developing Numeracy: Mental Maths

Year R supports the teaching of mental maths by providing a series of activities which develop these essential skills. On the whole the activities are designed for children to work on independently, although due to the young age of the children, the teacher may need to read the instructions with the children and ensure that they understand the activity before they begin working on it.

Year R develops concepts and skills for the different aspects of numeracy in the following ways:

An understanding of how numbers relate to each other
- counting forwards and backwards in ones;
- counting in twos and in tens;
- comparing and ordering numbers.

An understanding of place value
- counting to ten and beyond;
- counting in steps of ten.
(This helps the children begin to understand the role of digits.)

Recall of known facts
- beginning to recognise small numbers without counting;
- practising adding and subtracting.
(Over time, this will lead to some learning of simple facts.)

Mental calculation strategies
- beginning to relate addition to counting on;
- counting on from the larger number;
- counting on when one group of objects is hidden;
- beginning to relate the addition of doubles to counting on;
- selecting two groups of objects to make a given total;
- counting back from the larger number;
- counting up from the smaller number.

Extension

Many of the activity sheets end with a challenge (**Now try this!**) which reinforces and extends the children's learning, and provides the teacher with an opportunity for assessment. Again, it  may be necessary to read the instructions with the children before they attempt the activity. For some of the challenges the children will need to record their answers on a separate piece of paper.

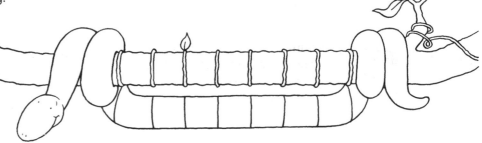

Organisation

Very little equipment is needed, but it will be useful to have available: coloured pencils, counters, dice, scissors, coins, number lines and number tracks.

To help teachers select appropriate learning experiences for the children, the activities are grouped into sections within the book. However, the activities are not expected to be used in this order; the sheets are intended to support, rather than direct, the teacher's planning.

Some activities can be made easier or more challenging by masking or substituting numbers. You may wish to re-use pages by copying them onto card and laminating them. If you find that the answer boxes are too small for the children's writing, you could enlarge the activity sheet onto A3 paper.

Teachers' notes

Brief notes are provided at the foot of each page giving ideas and suggestions for maximising the effectiveness of the activity sheets. These can be masked before copying.

Calculation strategies

Children in **Year R** may use the following strategies when working out additions:

Count all Children count each set in ones, then find the total by re-counting from one. For example, 2 + 5 is said as *one, two* followed by *one, two, three, four, five*; the total is then *one, two, three, four, five, six, seven*.

Count on Children count on from either the first number or the larger number to find the total. For example, 2 + 5 is said as *two, three, four, five, six, seven* (counting from the first number) or *five, six, seven* (counting from the larger number).

When working out subtractions, children may use the following strategies:

Count all Children count each set in ones. For example, 5 – 2 is said as *one, two, three, four, five* followed by what is being taken away: *one, two*. The remainder is then *one, two, three*.

Count back Children count back to find the answer. For example, 5 – 2 is said as *five, four, three*.

Count up Children count up to find the answer, e.g. 5 – 2 is said as *three, four, five; so I have counted up three*.

Whole-class warm-up activities

The following activities provide some practical ideas which can be used to introduce the main teaching part of the lesson.

Spot the mistake

This activity focuses on counting in order. Use a hand puppet or mitten to 'say' the numbers. Explain that the puppet sometimes makes mistakes, and the children should wave their hand if they spot one. Start counting from any small number and make a deliberate mistake: for example, *6, 7, 8, 10, 9…* Include number name errors: for example, *eleven, twelve, threeteen, fourteen…; eighteen, nineteen, tenteen…* Then do the same thing counting backwards from a larger number. You could also count forwards and backwards in twos or tens.

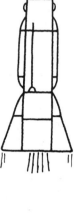

Counting round

Count forwards or backwards in ones around the class. This activity can also be used to practise counting in twos.

Counters

Give the children counters. Ask questions such as: *Pick up five counters. Pick up two more. How many counters have you now? Pick up six counters. Take away four. How many are left?*

What's the question?

Give the children the answer to a question (for example, 5) and ask: *Can you give me a question with the answer 5?* (for example, 4 + 1); *Can anyone give me another question with the answer 5?; Can you think of a take away question with the answer 5?* Discuss the range of calculations that give the answer 5.

Number cards

Give each child a set of 0 to 9 cards, which should be spread randomly face up on the table. Hold up a number card and ask: *Can you hold up the number that is one more/one less than the number on my card?*

Hold up a number card and ask: *Can you hold up two numbers that add to make the number on my card?* Explore the different sums that are produced.

Hold up two number cards. Ask: *Can you add the numbers on these cards together? Hold up the card that shows the answer.*

Hold up (or write on the board) a number in the range 11–20. Ask: *Can you hold up the number that is ten less than my number?*

Making stew

- **Read the rhyme.**
- **Write the numbers in the boxes.**

1, 2, make stew

3, 4, add more

5, 6, now mix

7, 8, dropped the plate!

9, 10, start again!

make stew

add more

now mix

dropped the plate!

start again!

Now try this!

- **Cut out the cards.**
- **Mix them up.**
- **Put them in order.**

Teachers' note At the start of the lesson, say the rhyme together several times using appropriate actions. You could then repeat the rhyme ten times, missing out numbers: the first time miss out the number one and put your finger on your lips, the second time miss out numbers one and two, and so on up to ten. Encourage the children to 'think' the missing numbers in their heads.

Developing Numeracy Mental Maths Year R © A & C BLACK

Tidy-up rhyme

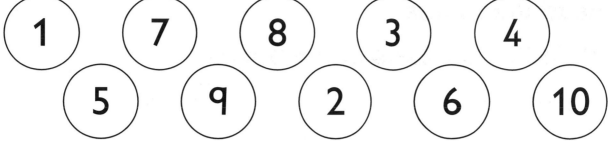

(1) (7) (8) (3) (4)

(5) (9) (2) (6) (10)

• **Fill in the numbers in the rhyme.**

(1) (2) so much to do!

( ) ( ) let's sweep the floor!

( ) ( ) pick up those bricks!

( ) ( ) wash up that plate!

( ) ( ) let's tidy again!

• **Colour the number** 2 **in the rhyme.**

Miss a number. Colour the next number.

Continue to the end.

Teachers' note At the start of the lesson, say the rhyme together using appropriate actions until the children are familiar with the words. When the children have completed the activity, you could ask them to help you change the words to make up a new rhyme: for example, '1, 2, so much to do! 3, 4, let's mop the floor! 5, 6, let's stack the bricks!' and so on.

**Developing Numeracy
Mental Maths Year R
© A & C BLACK**

This group is counting round the circle.

Miss Rix says (1) .   Jo says (2) .

Miss Rix

Jo

Ben

Amy

Lin

Tom

• **Who says these numbers?**

4	Tom
8	
3	

6	
5	
9	

Ben says 1 . Tom says 2 .

• **Who says these numbers?**

| 10 | |
| 20 | |

Now try this!

Teachers' note At the start of the lesson, sit in a circle with the children and count clockwise around the circle. Ask them to predict who will say a number, then test out their predictions. When completing the sheet, encourage the children to count around the circle in the picture, pointing to each child in turn.

Developing Numeracy Mental Maths Year R © A & C BLACK

Puppet mistakes

• **Spot the counting mistakes!**

Write the missing number.

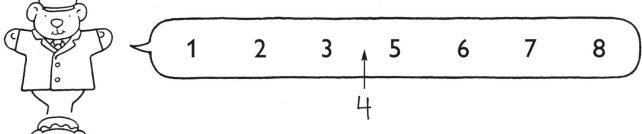

1   2   3   ↑   5   6   7   8

4

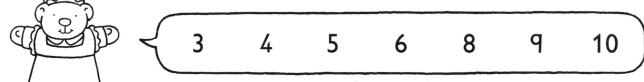

3   4   5   6   8   9   10

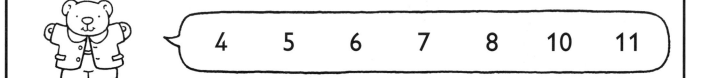

4   5   6   7   8   10   11

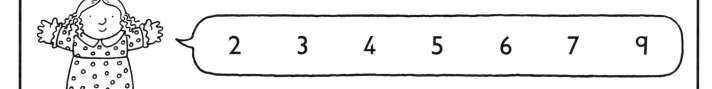

2   3   4   5   6   7   9

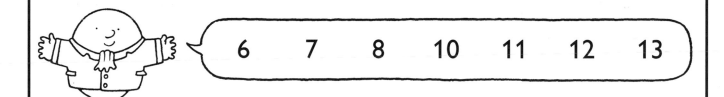

6   7   8   10   11   12   13

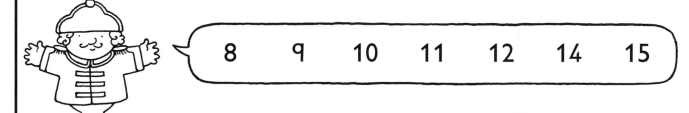

8   9   10   11   12   14   15

Now try this!

• **Write the numbers 1 to 15 in order.**

Teachers' note At the start of the lesson, play 'Spot the mistake' (see page 5). When completing the sheet, some children might find it helpful to refer to a number line.

**Developing Numeracy
Mental Maths Year R
© A & C BLACK**

Dot to dot

• **Count aloud. Join the dots in order.**

Start with 0 .

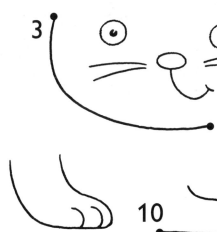

4
5
6
7
2
1
3
8
0
9
11
10

5
6
0
4
1
q
7
2
8 10
11
3

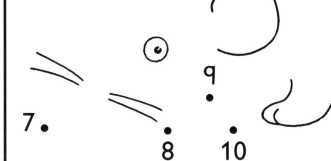
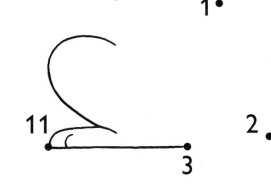

• Count on from 10. Write the numbers.

Now try this!

10
11

• **Join the dots in order.**

Teachers' note Make sure the children say the numbers aloud as they join the dots. If they are not confident with numerals to 10 and beyond, provide a number line or track. The children could cover each number on the line or track as they count along.

**Developing Numeracy
Mental Maths Year R
© A & C BLACK**

Up to bed

Each child steps up 2 stairs and ⟨counts on⟩ **2.**

• Fill in the missing numbers.

8, _____, _____

7, _____, _____

4, _____, _____

3, _____, _____

1, _____, _____

0, 1, 2

• Count on. Write the next 3 numbers.

2, _____, _____, _____   5, _____, _____, _____

11, _____, _____, _____   9, _____, _____, _____

Teachers' note To introduce the activity, mark lines a child's pace apart on the playground and number them from zero. Ask children to stand on a particular line and to count on two, three or four as they step forwards. When using the activity sheet, tell the children to start at the bottom of the stairs in the picture. Ask children who have stairs at home to count them as they climb, starting with zero.

Developing Numeracy
Mental Maths Year R
© A & C BLACK

Down to breakfast

Each child steps down 2 stairs and `counts back` 2.

• Fill in the missing numbers.

10, _9_, _8_

q, ____, ____

7, ____, ____

6, ____, ____

3, ____, ____

2, ____, ____

(Stair numbers, left side top to bottom: 10, q, 8, 7, 6, 5, 4, 3, 2, 1, 0)
(Stair numbers, right side top to bottom: 10, q, 8, 7, 6, 5, 4, 3, 2, 1, 0)

• Count back. Write the next 3 numbers.

8, _7_, ____, ____   11, ____, ____, ____

13, ____, ____, ____   20, ____, ____, ____

Teachers' note To introduce the activity, mark lines a child's pace apart on the playground and number them from zero. Ask children to stand on a particular line and to count back two, three or four as they step backwards. They could also count back as they move a toy along a number track. Encourage children who have stairs at home to count on and back as they go up and down.

Developing Numeracy
Mental Maths Year R
© A & C BLACK

12

Blast off!

• **Count down to** $\boxed{0}$.

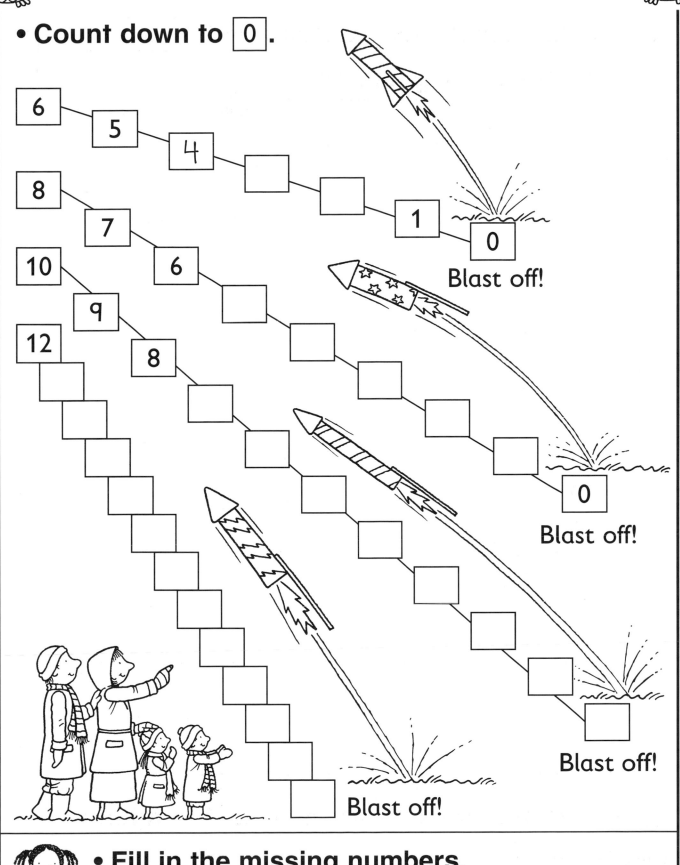

$\boxed{6}$ $\boxed{5}$ $\boxed{4}$ $\square$ $\square$ $\boxed{1}$ $\boxed{0}$ Blast off!

$\boxed{8}$ $\boxed{7}$ $\boxed{6}$ $\square$ $\square$ $\square$ $\square$ $\boxed{0}$ Blast off!

$\boxed{10}$ $\boxed{9}$ $\boxed{8}$ $\square$ $\square$ $\square$ $\square$ $\square$ $\square$ $\square$ Blast off!

$\boxed{12}$ $\square$ $\square$ $\square$ $\square$ $\square$ $\square$ $\square$ $\square$ $\square$ $\square$ $\square$ Blast off!

• **Fill in the missing numbers.**

18, 17, 16, 15, _____, _____, _____, _____

Teachers' note At the start of the lesson, ask the children to stand behind their chairs and crouch down. Begin at a suitable number and count down towards zero, encouraging the children to join in. When zero is reached, all shout 'Blast off!' and jump up into the air. Some children might need to refer to a number line when completing the activity sheet.

Developing Numeracy
Mental Maths Year R
© A & C BLACK

Space launch

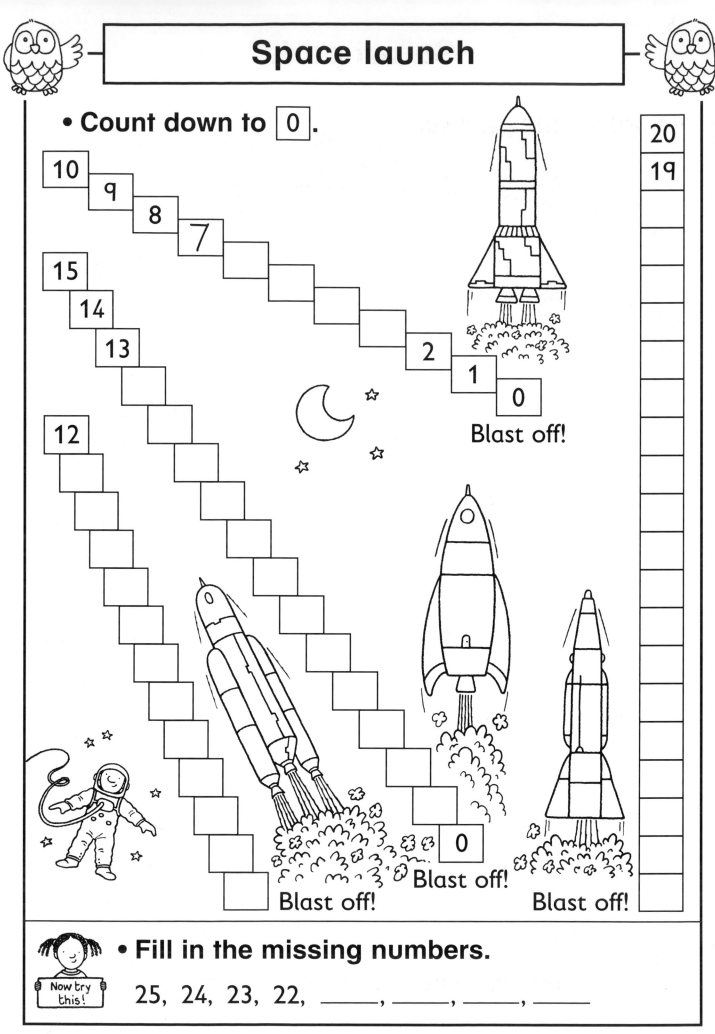

• **Count down to** $\boxed{0}$.

10
9
8
7

15
14
13

12

2
1
0

Blast off!

0

Blast off!

Blast off!

Blast off!

20
19

• **Fill in the missing numbers.**

25, 24, 23, 22, _____, _____, _____, _____

Teachers' note At the start of the lesson, ask the children to stand behind their chairs and crouch down. Begin at a suitable number and count down towards zero, encouraging the children to join in. When zero is reached, all shout 'Blast off!' and jump up into the air. Some children might need to refer to a number line when completing the activity sheet.

Developing Numeracy
Mental Maths Year R
© A & C BLACK

Safari park

- ## Look at the picture.

- ## Write how many

monkeys	1	elephants	
lions		zebras	
giraffes		rhinos	

- ## Draw 3 more cars on the road.
- ## How many cars are there now?

Teachers' note Where possible, assess whether the children are able to recognise the number of animals in a small group without pointing to each animal. If the children do point, encourage slow, careful counting to ensure that each animal is counted only once and none are missed. It can be helpful to place a counter or cube on each animal as it is counted.

Developing Numeracy Mental Maths Year R © A & C BLACK

Daisy, daisy

- **Count the daisies in each hoop.**
- **Count them again to check.**

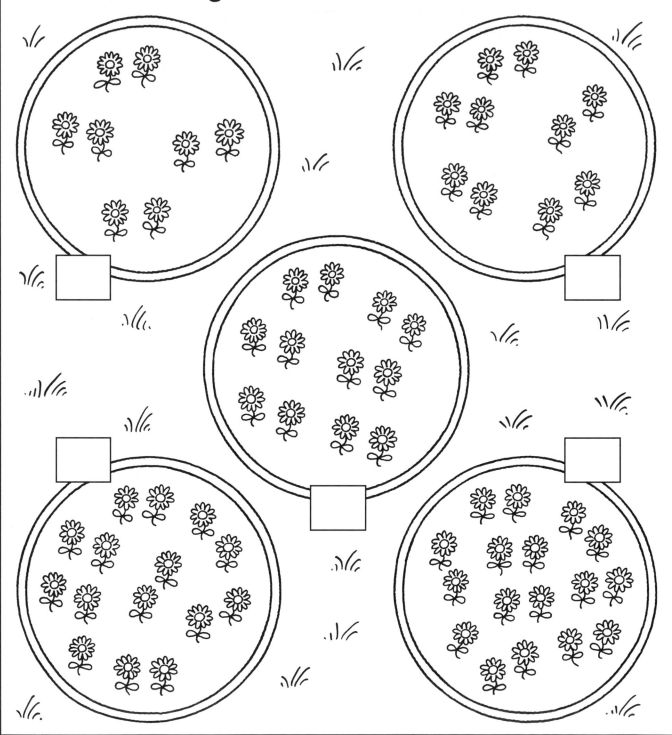

- **Draw rings around pairs of daisies.**
- **Now count the daisies in 2s.**

Teachers' note If there are daisies in the school grounds, you could try a similar activity outside. Give each group of children a small PE hoop. Ask them to place it on the ground and count the daisies in the hoop.

**Developing Numeracy
Mental Maths Year R
© A & C BLACK**

Action trail

- Take turns to move 1 place along the trail.
- Roll the dice. Do the action that number of times.

clap

start

jump

touch your ear

touch your nose

say 'hello'

hello

hop

open your mouth

touch your toes

stand up

turn around

smile

finish

wrinkle your nose

Teachers' note Photocopy this page onto A3. Give each group one copy of the page and a dice with *numerals* from 1 to 6 (or, for more confident children, other numerals up to 10). Each child will need a different-coloured counter. Explain that the game is not a race to the finish. While one child performs the actions to match the number on the dice, the others in the group should count aloud.

Developing Numeracy Mental Maths Year R © A & C BLACK

Counting fun

This group is counting in 10s round the circle.

Mr Ross says (10) . **Kim says** (20) .

Mr Ross

Kim

Ali

Sam

Meg

Fay

• **Who says these numbers?**

(30)⊃ | Sam | (60)⊃ [ ]

(80)⊃ [ ]   (50)⊃ [ ]

(40)⊃ [ ]   (90)⊃ [ ]

Now try this!

Sam says [10]. **Fay says** [20].

• **Who says these numbers?**

(70)⊃ [ ]   (100)⊃ [ ]

Teachers' note At the start of the lesson, sit in a circle with the children and count in tens clockwise around the circle. Ask them to predict who will say a number, then test out their predictions. When completing the sheet, encourage the children to count around the circle in the picture, pointing to each child in turn.

Developing Numeracy Mental Maths Year R © A & C BLACK

How much money?

- **Point to each coin and count in 10s.**

- **Write how much money is in each row.**

_____ P

_____ P

_____ P

_____ P

_____ P

- **Cut out the cards.**

- **Put them in order.**

Start with the ⬚smallest⬚ **amount.**

Teachers' note Practise counting in tens at the start of the lesson. Ensure that the children can point accurately to items (such as 10p coins) when saying the numbers aloud. Some children might find it helpful to have actual 10p coins placed on the coin pictures. Further questions can be asked, such as, 'What if we took away 10p from each row?' 'What if we had an extra 10p in each row?'

**Developing Numeracy
Mental Maths Year R
© A & C BLACK**

• **Write the numbers in order on the frogs.**

• **Colour frog** 2 **green. Miss a frog.**

Colour the next frog. Continue to the end.

• **Write the numbers from the green frogs.**

Now try this!

2   4

• **Say the numbers.**

Teachers' note This page helps the children to realise that listing every other number when counting in ones is the same as counting in twos. Begin the lesson by counting in ones, whispering and shouting alternate words, i.e. 'one, TWO, three, FOUR…' Progress to saying only every other word: 'two, four, six…' Explain that this is known as counting in twos.

**Developing Numeracy
Mental Maths Year R
© A & C BLACK**

Two by two

- • **Point to each coin and count in 2s.**
- • **Write how much money is in each row.**

_____ P

_____ P

_____ P

_____ P

_____ P

- • **Cut out the cards.**
- • **Put them in order.**

Start with the smallest **amount.**

Teachers' note Practise counting in twos at the start of the lesson using the 'whisper/shout' approach (see page 20). You could also use the 'Counting round' activity on page 5. Some children might find it helpful to have actual 2p coins placed on the coin pictures. Further questions can be asked, such as, 'What if we took away 2p from each row?' 'What if we had an extra 2p in each row?'

Developing Numeracy
Mental Maths Year R
© A & C BLACK

Birds on eggs

- **Read how many eggs are in each nest.**
- **On each branch, colour the nest with** more **eggs.**

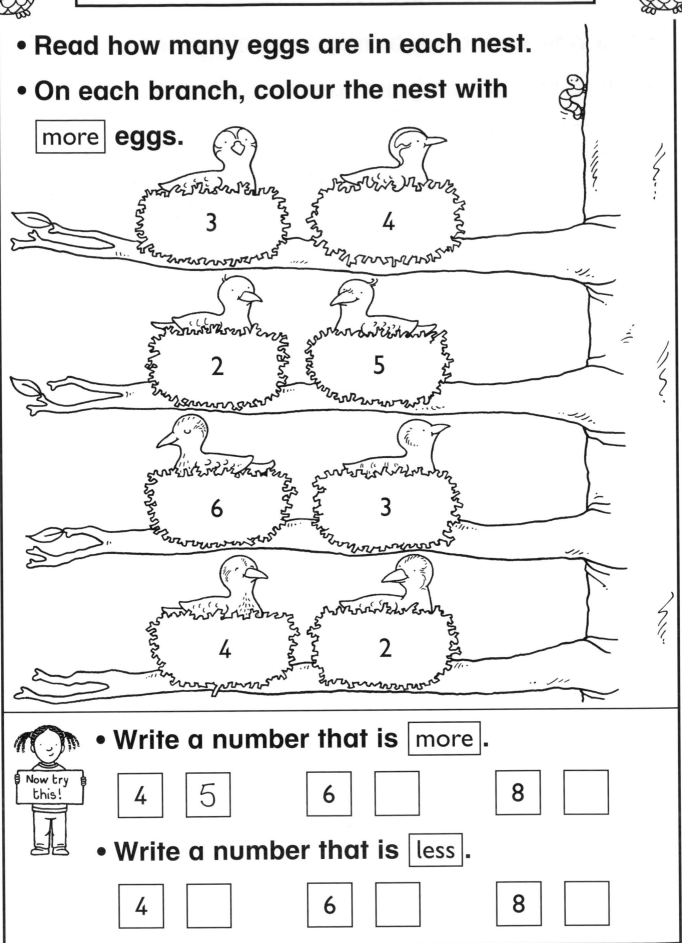

- **Write a number that is** more **.**

| 4 | 5 | | 6 | | | 8 | |

- **Write a number that is** less **.**

| 4 | | | 6 | | | 8 | |

Now try this!

Teachers' note This activity encourages the children to compare the size of numbers without counting objects. Children who find this difficult could draw the eggs next to each nest and compare the groups of eggs. Alternatively, they could find both numbers on a number line and see which is further left or right.

**Developing Numeracy
Mental Maths Year R
© A & C BLACK**

Naughty Felix

Felix has torn a tape measure into pieces.

• Fill in the missing numbers.

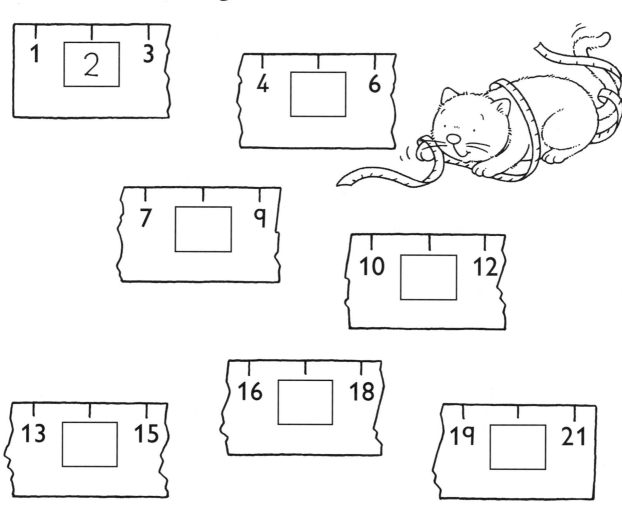

• Fill in the missing numbers.

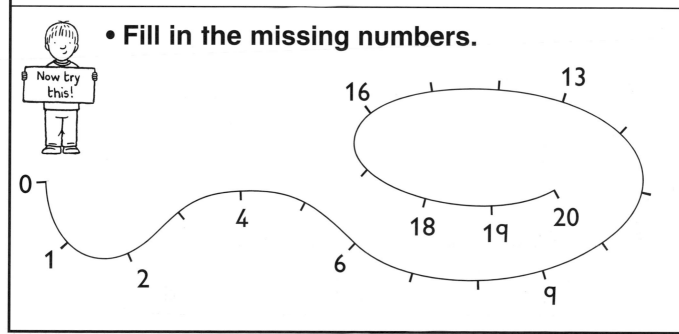

Teachers' note At the start of the lesson, call out two numbers (or hold up two number cards) and ask the children to give a number that lies between the two. If more than one number lies between the given numbers, discuss which numbers these are and in what order they come.

**Developing Numeracy
Mental Maths Year R
© A & C BLACK**

Ella's number

- ## Write Ella's number in the box.

 5 | Ella's number is **more than** 5.  Ella's number is **less than** 7. 7

 7 | Ella's number is **less than** 7.  Ella's number is **greater than** 5. 5

 8 | Ella's number is **smaller than** 8.  Ella's number is **larger than** 6. 6

 9 | Ella's number is **more than** 9.  Ella's number is **less than** 11. 11

- ## Write a number that lies [between] these.

Now try this!

10 [ ] 13   11 [ ] 14

13 [ ] 16   15 [ ] 19

Teachers' note During the first part of the lesson, encourage the children to make up their own puzzles using number cards: for example, 'My number is more than Liam's number but less than Jamal's'. In the extension activity, discuss that more than one number lies between the two given. Provide number lines or tracks if the children need further support.

**Developing Numeracy
Mental Maths Year R
© A & C BLACK**

24

Crunchy cookies

- **Write how many chocolate drops are on each cookie.**

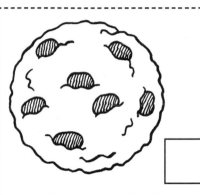 3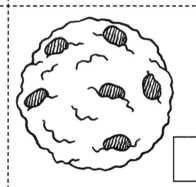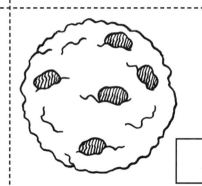

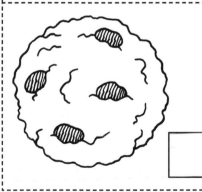 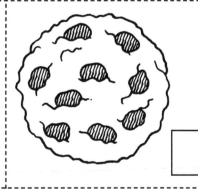 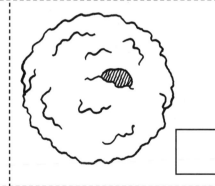

- **Cut out the cards.**

- **Pick 4 cards. Put them in order. Start with the cookie with the** fewest **chocolate drops.**

- **Put all the cards in order.**

Start with the cookie with the fewest drops.

Teachers' note The children could work with a partner and keep picking different sets of four cards to order. Encourage more confident children to try five or six cards at a time. They could write down the numbers in order on a piece of paper each time.

**Developing Numeracy
Mental Maths Year R
© A & C BLACK**

Number snakes

- **Point to the numbers and say them.**

- **Write them in order on the snake.**

  Start with the | smallest | **number.**

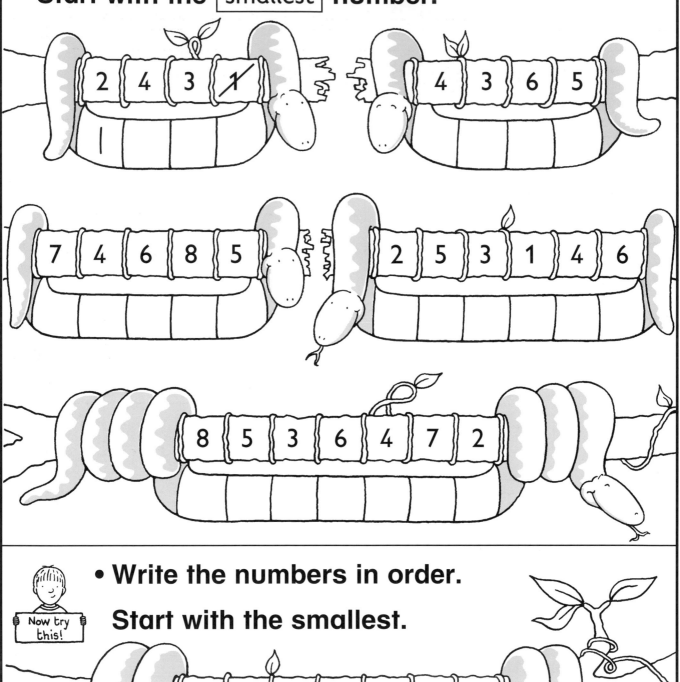

- **Write the numbers in order.**

  Start with the smallest.

Now try this!

Teachers' note Encourage the children to cross off each number as they copy it onto the snake. For further practice in ordering a given set of numbers, the children could shuffle number cards and order them.

Developing Numeracy Mental Maths Year R © A & C BLACK

Fruit and veg

• **Colour the** 3rd **pumpkin orange.**

1st

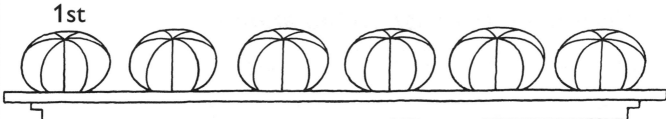

• **Colour the** 5th **and** 7th **tomatoes red.**

1st

• **Colour the** 2nd , 4th **and** 6th **cabbages green.**

1st

1st

Now try this!

• **Draw the fruit that is**

3rd	5th
2nd	4th
6th	7th

Teachers' note At the start of the lesson, write this list on the board: 1st, 2nd, 3rd, 4th, 5th, 6th, 7th, 8th. Discuss what these mean and ask the children to say them aloud. Invite several children to form a queue and to call out who is first, second, third, and so on. You could also show a selection of items in a line and ask the children to identify which is third, fifth, second, and so on.

Developing Numeracy
Mental Maths Year R
© A & C BLACK

Spotty dogs

- **Draw** 1 more **spot on each dog.**
- **How many spots are there now?**

5 1 more ⟩ 6

6 1 more ⟩ ☐

8 1 more ⟩ ☐

4 1 more ⟩ ☐

7 1 more ⟩ ☐

9 1 more ⟩ ☐

 Now try this!

- **Fill in the missing numbers.**

5 1 more ⟩ ☐ 1 more ⟩ ☐ 1 more ⟩ ☐

9 1 more ⟩ ☐ 1 more ⟩ ☐ 1 more ⟩ ☐

Teachers' note Reinforce 'one more' using the 'Number cards' activity on page 5. Remind the children that 'one more' is the next number they come to when counting on. After completing the activity sheet, ask the children to say their answers aloud: for example, '6 is one more than 5.' Introduce other words that can be used, such as 'and one' or 'add one'.

**Developing Numeracy
Mental Maths Year R
© A & C BLACK**

Sizzling sausages

• **If** 1 **sausage falls off,**

how many are left?

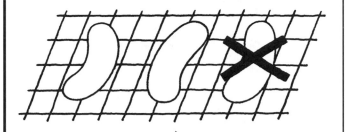

3 | 1 less ➤ | 2

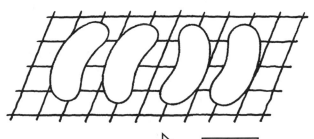

4 | 1 less ➤ |

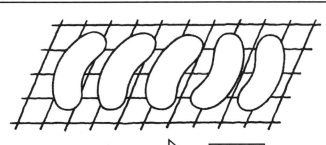

5 | 1 less ➤ |

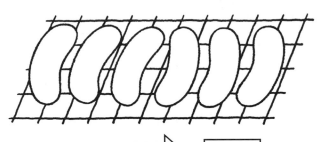

6 | 1 less ➤ |

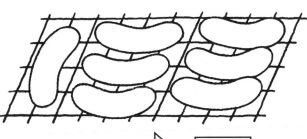

7 | 1 less ➤ |

8 | 1 less ➤ |

Now try this!

• **Fill in the missing numbers.**

9 | 1 less ➤ | | 1 less ➤ | | 1 less ➤ |

5 | 1 less ➤ | | 1 less ➤ | | 1 less ➤ |

Teachers' note Reinforce 'one less' using the 'Number cards' activity on page 5. Remind the children that 'one less' is the next number they come to when counting back. After completing the activity sheet, ask the children to say their answers aloud: for example, '2 is one less than 3.' Introduce other words that can be used, such as 'take one' or 'take away one'.

**Developing Numeracy
Mental Maths Year R
© A & C BLACK**

Jack and the bean pods

- ## Write how many beans are in each hand.

- ## Write how many | altogether | .

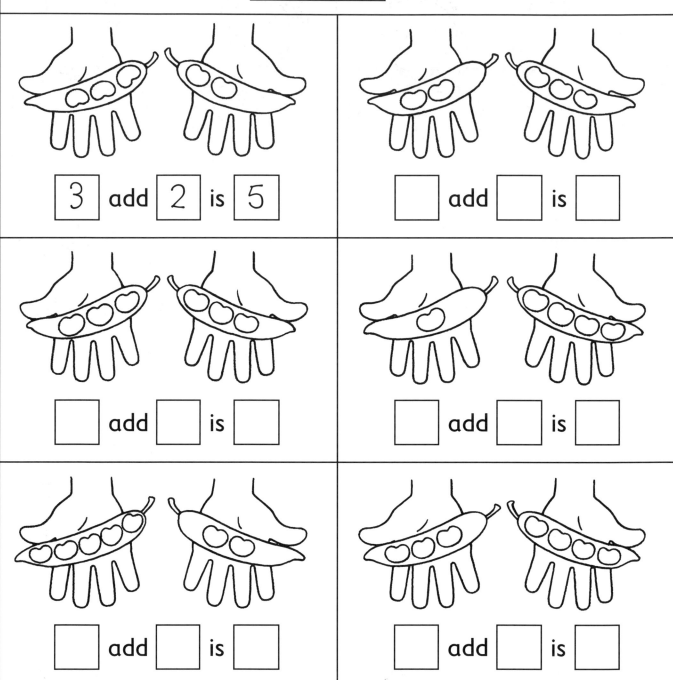

| 3 | add | 2 | is | 5 |

[ ] add [ ] is [ ]

[ ] add [ ] is [ ]

[ ] add [ ] is [ ]

[ ] add [ ] is [ ]

[ ] add [ ] is [ ]

Now try this!

- ## Draw some beans. Write the addition.

[ ] add [ ] is [ ]

Teachers' note You could begin the lesson with the 'Counters' activity on page 5. This can be linked with the story of *Jack and the Beanstalk*. Ask, 'If Jack has three beans in one hand and two in the other, how many does he have altogether?' The children could use counters to represent the beans. Discuss other ways of solving the additions, such as counting on using a number line.

Developing Numeracy
Mental Maths Year R
© A & C BLACK

At the sweet shop

- **Write how many pennies each child spends.**

- **How many pennies does this child spend?**

Teachers' note The children may require 1p coins to help them answer these questions. Discuss other ways of solving the additions, such as counting on from the first number, or counting on from the larger number. Provide number lines or tracks if appropriate. Encourage the children to say the questions and answers aloud to reinforce addition vocabulary: for example, '1p and 1p makes 2p.'

Developing Numeracy
Mental Maths Year R
© **A & C BLACK**

Great grapes

0 1 2 3 4 5 6 7 8

- **Count on from the first number.**

- **How many grapes altogether?**

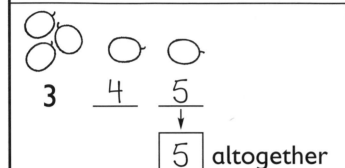

3   _4_   _5_
        ↓
      [5] altogether

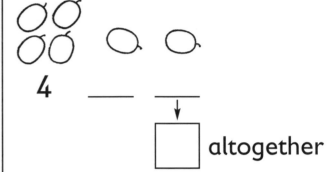

4   ___
        ↓
      [ ] altogether

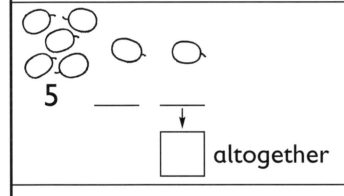

5   ___
        ↓
      [ ] altogether

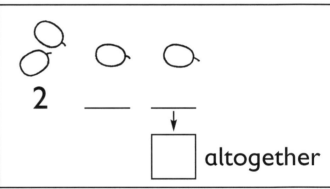

2   ___
        ↓
      [ ] altogether

- **How many grapes altogether?**

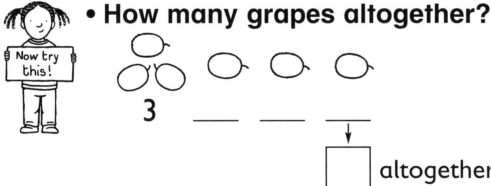

3   ___   ___   ___
              ↓
            [ ] altogether

4   ___   ___   ___
              ↓
            [ ] altogether

Teachers' note Suggest that the children touch each grape in turn as they count on. (Watch out for those who count on incorrectly, beginning on the number itself.) Encourage the children to say the questions and answers aloud to reinforce addition vocabulary: for example, 'We started with three grapes and counted on two more to get five grapes,' or 'Three add two makes five.'

Developing Numeracy
Mental Maths Year R
© A & C BLACK

A head start

- **Put the number in your head.**
- **Count on** $\boxed{3}$ **.**

- **Put the number in your head.**
- **Count on** $\boxed{5}$ **.**

Teachers' note Some children may find it helpful to touch their head and say the starting number aloud before counting on using their fingers. If necessary, provide number lines to help the children remember which numbers come next. Watch out for children who count on incorrectly, beginning on the number itself: for example, (*Count on 3 from 4*) '4, 5, 6'.

Developing Numeracy
Mental Maths Year R
© A & C BLACK

Mirror, mirror

- **Draw** the same number **of faces in the mirror.**

- **Write how many altogether.**

double 3

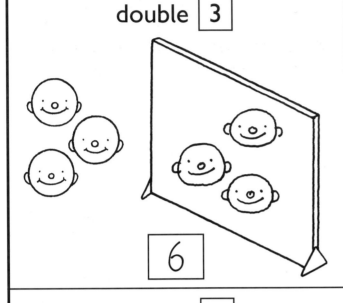

6

double 2

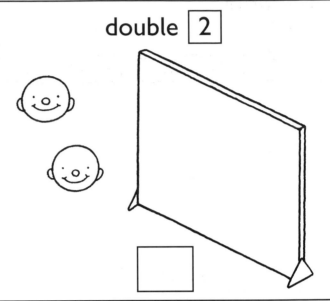

double 4

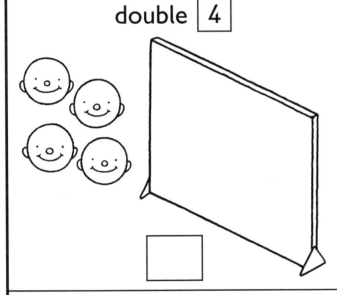

double 5

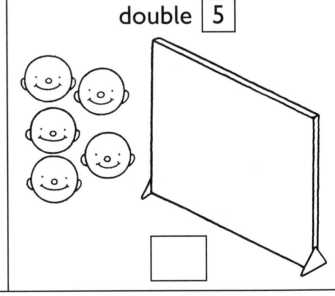

Now try this!

- **Write the** doubles .

double 4 is 8

double 3 is

double 2 is

double 1 is

double 5 is

double 6 is

Teachers' note Introduce or revise 'double'. When the children have completed the activity, ask them similar questions using a range of vocabulary: for example, 'What is twice three?', 'Two times two is…', 'Two lots of four is…', 'Five add five makes…'

Developing Numeracy
Mental Maths Year R
© A & C BLACK

Cats and mice

• **Put the first number in your head and count on.**

mice in
the hole

| 3 | 🐭 4 🐭 5 | 5 | altogether |

| 4 | 🐭 🐭 🐭 | ☐ | altogether |

| 5 | 🐭 🐭 | ☐ | altogether |

| 3 | 🐭 🐭 🐭 | ☐ | altogether |

| 5 | 🐭 🐭 🐭 | ☐ | altogether |

Now try this!

• **Put the** larger **number in your head and count on.**

🐱 🐱 and 🧺 6 is ☐

🐱 🐱 🐱 and 🧺 7 is ☐

🐱 🐱 🐱 🐱 and 🧺 6 is ☐

Teachers' note The children may find it helpful to touch their head and say the starting number aloud before counting on as they point to the mice or kittens.

**Developing Numeracy
Mental Maths Year R
© A & C BLACK**

35

Busy buses

- **Read how many people are on the bus.**

- **Count on. Write how many people altogether.**

5   6 7 8

_8_ people altogether

4   ___ people altogether

4   ___ people altogether

3   ___ people altogether

4   ___ people altogether

5   ___ people altogether

- **Make up your own bus question.**

Now try this!

___ people altogether

Teachers' note Watch out for children who count on incorrectly, beginning on the number itself: for example, (*Count on 3 from 5*) '5, 6, 7'. If necessary, provide number lines to help the children remember which numbers come next.

**Developing Numeracy
Mental Maths Year R
© A & C BLACK**

Yoghurt splits

Each pack has 6 yoghurts.

The pack splits into 2 parts.

• 6 splits into

 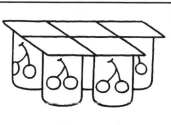

 4 and 2 | | and | |

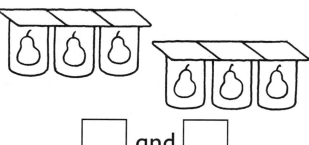 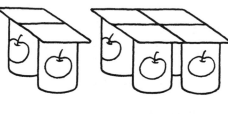

| | and | | | | and | |

| | and | | | | and | |

• Draw 6 dots on each domino.

Make all the dominoes different.

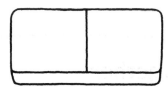 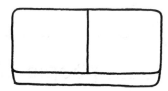 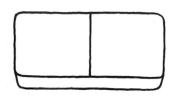

Teachers' note To help the children begin to learn these number facts, encourage them to say aloud how the number 6 can be split into two parts. Some children may need to look at a set of dominoes for the extension activity.

**Developing Numeracy
Mental Maths Year R
© A & C BLACK**

Seven little fish

- Draw ⬛7 fish in 2 tanks.

- Write the addition sentence.

| 4 | and | 3 | is 7 |

| | and | | is 7 |

| | and | | is 7 |

| | and | | is 7 |

| | and | | is 7 |

| | and | | is 7 |

- **Fill in the missing numbers.**

Now try this!

3 and ⬜ makes 7   5 add ⬜ is 7

1 add ⬜ is 7   0 and 7 makes ⬜

Teachers' note Ensure the children realise that they must draw a total of seven fish in the two tanks (and that all seven fish could be in one of the tanks). They should make sure each picture gives a different addition sentence. To help the children begin to learn these number facts, encourage them to say aloud how the number 7 can be split into two parts.

**Developing Numeracy
Mental Maths Year R
© A & C BLACK**

Elephant parade

- **Colour each row of elephants in a different way.**

  Use red and blue .

- **Write the addition sentence.**

red		blue	altogether
4	and	1	= 5

☐ and ☐ = ☐

☐ and ☐ = ☐

☐ and ☐ = ☐

☐ and ☐ = ☐

☐ and ☐ = ☐

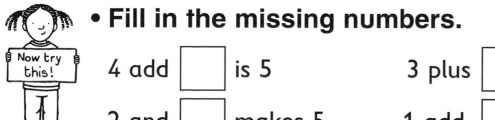

- **Fill in the missing numbers.**

4 add ☐ is 5   3 plus ☐ equals 5

2 and ☐ makes 5   1 add ☐ is 5

Teachers' note Ensure the children understand that the elephants in a row can all be the same colour (i.e. all red or all blue). When they have completed the activity, discuss the different ways of describing the number facts shown in the extension activity. Help the children begin to learn some of these facts by asking oral questions, using a range of vocabulary.

**Developing Numeracy
Mental Maths Year R
© A & C BLACK**

Tea and biscuits

- ## Write how many cups are on each tray.

4

- ## Which trays have a [total] of [8]? Join them.

- ## Join plates that have a total of [10].

6

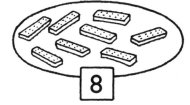
8

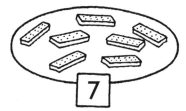
7

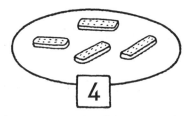
4

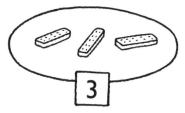
3

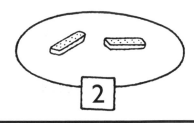
2

Teachers' note Revise 'total' if necessary. This activity helps the children begin to learn addition
facts by heart by recognising the totals of pairs of numbers. Ask them to check their answers by
counting all the cups on the trays they have joined. You could mask some of the pictures to
differentiate the activity.

**Developing Numeracy
Mental Maths Year R
© A & C BLACK**

40

Milk teeth

- **How many teeth are left?**

I had 6 teeth. 2 fell out.

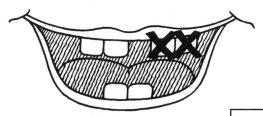

6 take away 2 = 4

I had 5 teeth. 3 fell out.

5 take away 3 =

I had 7 teeth. 1 fell out.

7 take away 1 =

I had 6 teeth. 3 fell out.

6 take away 3 =

I had 5 teeth. 2 fell out.

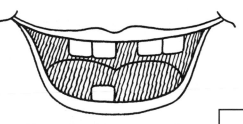

5 take away 2 =

I had 8 teeth. 2 fell out.

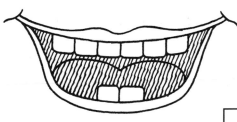

8 take away 2 =

Now try this!

- **Tick ✔ the questions that have the answer 1 .**

4 take away 3 ☐   6 take away 4 ☐

7 take away 6 ☐   5 take away 4 ☐

Teachers' note Demonstrate how to cross off the teeth that have fallen out and to count how many are left. Some children may find it helpful to use counters to represent the teeth. Discuss other ways of solving these questions, such as counting back from the larger number.

Developing Numeracy Mental Maths Year R © A & C BLACK

Pick and mix

• **How many sweets are left?**

I had ⟨5⟩ sweets. I ate ⟨2⟩.

5 take away 2 = ⟨3⟩

I had ⟨4⟩ sweets. I ate ⟨3⟩.

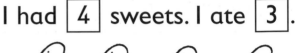
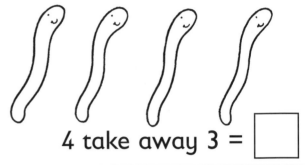

4 take away 3 = ☐

I had ⟨6⟩ sweets. I ate ⟨3⟩.

6 take away 3 = ☐

I had ⟨6⟩ sweets. I ate ⟨4⟩.

6 take away 4 = ☐

I had ⟨7⟩ sweets. I ate ⟨4⟩.

7 take away 4 = ☐

I had ⟨5⟩ sweets. I ate ⟨5⟩.

5 take away 5 = ☐

Now try this!

• **Tick ✔ the questions that have the answer ⟨3⟩.**

4 take away 1 ☐   5 take away 3 ☐

7 take away 5 ☐   8 take away 5 ☐

Teachers' note Demonstrate how to cross off the sweets that have been eaten and to count how many are left. Some children may find it helpful to use counters to represent the sweets. Discuss other ways of solving these questions, such as counting back from the larger number.

Developing Numeracy Mental Maths Year R © A & C BLACK

Take it away

• **Put the first number in your head.**

• Count back **3.**

7 take away 3 = $\boxed{4}$

5 take away 3 = $\boxed{}$

6 take away 3 = $\boxed{}$

8 take away 3 = $\boxed{}$

4 take away 3 = $\boxed{}$

9 take away 3 = $\boxed{}$

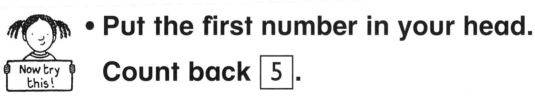 • **Put the first number in your head.**

Count back $\boxed{5}$.

10 take away 5 = $\boxed{}$

12 take away 5 = $\boxed{}$

Teachers' note Revise counting back at the start of the lesson. Some children may find it helpful to touch their head and say the starting number aloud before counting back using their fingers. If necessary, provide number lines to help the children remember which numbers come next.

Developing Numeracy
Mental Maths Year R
© A & C BLACK

43

Magic number 7

There are 7 rabbits in a hat. Some jump out.

• **How many rabbits are left in the hat?**

| Count back | **from** | 7 |.

2 jump out

5 left in the hat

4 jump out

☐ left in the hat

3 jump out

☐ left in the hat

1 jumps out

☐ left in the hat

5 jump out

☐ left in the hat

6 jump out

☐ left in the hat

• **Count back from the larger number.**

6 take away 2 is ☐   8 take away 2 is ☐

9 take away 4 is ☐   10 take away 3 is ☐

Teachers' note When counting back from the larger number, the children will need to take care to count back the correct number. Encourage them to point to the rabbits or cross them off as they count back. Alternatively, they could use their fingers to keep count.

**Developing Numeracy
Mental Maths Year R
© A & C BLACK**

Little kittens

6 little kittens went out one day.

4 came home.

2 ran away.

• **How many kittens ran away?**

Count up **from the smaller number.**

5	went out		6	went out
2	came home		3	came home
	ran away			ran away
3	went out		5	went out
1	came home		3	came home
	ran away			ran away
5	went out		4	went out
1	came home		1	came home
	ran away			ran away

• **Make up your own kitten question.**

• **Give it to a friend to answer.**

Teachers' note For this activity the children should use the counting up method, rather than counting all or counting back. At the start of the lesson, say the rhyme together several times and explain how to count up from the number of kittens that came home: 'Four kittens came home, so we count up from four to find out how many ran away.' Some children may need a number line or track.

**Developing Numeracy
Mental Maths Year R
© A & C BLACK**

Music time

Each child needs an instrument.

- **How many** | more | **instruments are needed?**

  Count up from the smaller number.

| 3 | instruments | 5 | children | 2 | more needed |

| 2 | instruments | 6 | children | | more needed |

| 1 | instrument | 5 | children | | more needed |

| 2 | instruments | 4 | children | | more needed |

- **Count up from the smaller number.**

 7 take away 4 is [ ]    8 take away 5 is [ ]

10 take away 8 is [ ]   10 take away 7 is [ ]

Teachers' note For this activity the children should use the counting up method, rather than counting all or counting back. At the start of the lesson, ask several children to stand at the front, then show some musical instruments and ask how many more are needed for the children to have one each. Some children may find it helpful to use a number line or track when completing the sheet.

Developing Numeracy
Mental Maths Year R
© A & C BLACK

Teachers' note These cards can be enlarged, coloured and cut out for use in the home corner. Encourage discussion of the number of items and invite the children to compare them using questions such as, 'Which card has one more banana/one fewer bun than this one?' The cards can also be used in shopping role play situations (for example, where each bun costs 1p or 10p).

**Developing Numeracy
Mental Maths Year R
© A & C BLACK**

Teachers' note These cards can be enlarged, coloured and cut out for use in the home corner. Encourage discussion of the number of items and invite the children to compare them using questions such as, 'How many more gingerbread people are on this card than on this one?' The cards can also be used in shopping role play situations (for example, where each gingerbread person costs 1p or 10p).

**Developing Numeracy
Mental Maths Year R
© A & C BLACK**